Meet Me on the Farm

Learning the Long E Sound

Shelby Braidich

Rosen Classroom Books and Materials™
New York

Meet me on the farm.

We see a barn.

I feed a cow.

We feed a pig.

We feed a sheep.

We see a tractor.

We ride horses.

We plant seeds.

17

We pull weeds.

We lay on the green grass.

Word List

feed

green

me

meet

see

seeds

sheep

we

weeds

Instructional Guide

Note to Instructors:
One of the essential skills that enable a young child to read is the ability to associate letter-sound symbols and blend these sounds to form words. Phonics instruction can teach children a system that will help them decode unfamiliar words and, in turn, enhance their word-recognition skills. We offer a phonics-based series of books that are easy to read and understand. Each book pairs words and pictures that reinforce specific phonetic sounds in a logical sequence. Topics are based on curriculum goals appropriate for early readers in the areas of science, social studies, and health.

Letter/Sound: long e – To review the **short e** vowel sound, write the following words on a chalkboard or dry-erase board: *bed, set, red, fell, tell, Ted, Ed*. Have the child decode each word and underline the **short e**. Write and pronounce the following **long e** words: *deep, feed, keep, meet, beet, see, be, me*. Have the child tell how these words both look and sound different from the **short e** words. Ask them to underline **e** or **ee** in each word.

Phonics Activities: Have the child name the **long e** words they hear in the following oral sentences: "We play hide and seek every week." "The heel of my sock has a teeny tiny hole in it." "It's time to go to sleep." "Pete meets his friends at the park." "Close your eyes and don't peek." "These weeds are making me sneeze." As the child responds, list the **long e** words on the chalkboard or dry-erase board.
- Write the sentences from the previous activity on chart paper. Have the child underline the **long e** words. Make matching sentence strips and have the child match them to the sentences on the chart paper.
- Write the following words on the chalkboard or dry-erase board: *bed, seem, set, sell, deed, sleep, me, met, mess, meet, less, seed, sheep, get, green, she, feel, feet, fell, weeds, steep*, etc. Read each word and have the child indicate if the **e** in each word is long or short by clapping once for **short e** words and twice for **long e** words.

Additional Resources:
- Coster, Patience. *Farming & Industry*. Danbury, CT: Children's Press, 1998.
- DK Publishing Staff. *A Day at Greenhill Farm: Level 1*. New York: DK Publishing, Inc., 1998.
- Fowler, Allan. *Living on a Farm*. Danbury, CT: Children's Press, 2000.
- Gibbons, Gail. *Farming*. New York: Holiday House, 1988.

Published in 2002 by The Rosen Publishing Group, Inc.
29 East 21st Street, New York, NY 10010

Copyright © 2002 by The Rosen Publishing Group, Inc.

All rights reserved. No part of this book may be reproduced in any form without permission in writing from the publisher, except by a reviewer.

Book Design: Ron A. Churley

Photo Credits: Cover © Ben Mitchell/Image Bank; p. 3 © SuperStock; p. 5 © Dave Rusk/Index Stock; p. 7 © Inga Spence/Index Stock; p. 9 © Steve Satushek/Image Bank; p. 11 © Robin Schwartz/International Stock; pp. 13, 17 © Digital Stock; p. 15 © Don Stevenson/Index Stock; p. 19 © Bob Daemmrich/Stock, Boston/Picture Quest; p. 21 © Daniel Fort/Index Stock.

Braidich, Shelby, 1971-
   Meet me on the farm : learning the long E sound / Shelby Braidich. — 1st ed.
     p. cm. — (Power phonics/phonics for the real world)
   ISBN 0-8239-5920-1 (lib. bdg. : alk paper)
   ISBN 0-8239-8265-3 (pbk. : alk. paper)
   6-pack ISBN 0-8239-9233-0
   1. Farms—Juvenile literature. 2. English language—Vowels—Juvenile literature. [1. Farms.] I. Title. II. Series.
S519.B77 2001
630—dc21

                                2001000190

Manufactured in the United States of America